BUSINESS AND COMMERCIAL ARITHMETIC

ARITHMETIC FOR LIVING

BUSINESS AND COMMERCIAL ARITHMETIC

D.U.STEELE

CAMBRIDGE UNIVERSITY PRESS
Cambridge
London New York New Rochelle
Melbourne Sydney

Published by the Press Syndicate of the University of Cambridge
The Pitt Building, Trumpington Street, Cambridge CB2 1RP
32 East 57th Street, New York, NY 10022, USA
296 Beaconsfield Parade, Middle Park, Melbourne 3206, Australia

© Cambridge University Press 1980

First published 1980

Printed in Great Britain by
Spottiswoode Ballantyne Ltd
Colchester and London

Typeset by
Reproduction Drawings Ltd, Sutton, Surrey

ISBN 0 521 29924 1

Contents

Introduction

The Arithmetic for Living series consists of an integrated course that is aimed mainly at pupils aged between 13 and 16 years. The books are also suitable for tertiary students who have left school without attaining the minimum competency that is required for courses leading to qualifications in craft and technical studies.

There are two main books: *Foundation Arithmetic* and *Basic Arithmetic*. In addition, there are three Topic Books: *Business Arithmetic, Home Arithmetic* and *Everyday Statistics*. All these books can be used independently of one another or linked to any scheme which a class may be following. While the emphasis is on metric units, there is an adequate treatment of all imperial units which continue to be used in industry and other employment, everyday life and examinations.

The exercises are carefully graded. Sometimes they are preceded by some key facts and/or an example to illustrate how answers and working should be presented. Whenever practical the first question in a line is typical of those that follow. Answers to all problems are provided.

1 Interest

Exercise 1. Simple interest

1 Find the simple interest payable on each of the loans listed below.

	Principal £	Rate % per annum	Time in years
a	1 300	2	4
b	7 500	5	3
c	600	8	7
d	4 200	6	2
e	10 400	10	3
f	1 250	$3\frac{1}{2}$	6
g	220	$7\frac{1}{2}$	$3\frac{1}{2}$
h	2 220	$11\frac{1}{2}$	$4\frac{1}{2}$
i	6 480	$12\frac{1}{2}$	$2\frac{1}{2}$
j	25 500	$6\frac{1}{2}$	$6\frac{1}{2}$
k	6 360	7	$2\frac{1}{4}$
l	3 480	$8\frac{1}{4}$	$3\frac{3}{4}$
m	750	9	4
n	640	$9\frac{1}{4}$	4
o	960	$13\frac{1}{2}$	$3\frac{1}{2}$
p	4 750	$8\frac{3}{4}$	$5\frac{1}{2}$
q	8 300	$7\frac{1}{4}$	$2\frac{3}{4}$
r	3 230	$6\frac{1}{2}$	5
s	12 250	$7\frac{1}{2}$	$6\frac{3}{4}$
t	10 310	$5\frac{1}{2}$	$3\frac{1}{2}$

2 Find the rate of interest charged on each of the loans listed below.

	Principal £	Interest £	Time in years
a	3 000	1050·00	5
b	5 500	907·50	3
c	12 000	2475·00	$2\frac{1}{2}$
d	3 750	2587·50	6
e	1 550	627·75	$4\frac{1}{2}$
f	24 000	4860·00	$1\frac{1}{2}$

3 Find the time in years for each of the loans listed below.

	Principal £	Interest £	Rate % per annum
a	6 500	1300·00	5
b	2 800	168·00	2
c	4 750	1923·75	9
d	21 000	4331·25	$7\frac{1}{2}$
e	625	287·50	$11\frac{1}{2}$
f	1 780	589·63	$13\frac{1}{4}$

4 Find the principal which earns the interest in the loans listed below.

	Interest £	Time in years	Rate % per annum
a	450·00	4	$7\frac{1}{2}$
b	1800·00	$1\frac{1}{2}$	6
c	864·00	3	9
d	143·00	2	11
e	520·63	$3\frac{1}{2}$	$8\frac{1}{2}$
f	284·63	$2\frac{3}{4}$	$11\frac{1}{2}$

Exercise 2. Simple interest

1 Find the simple interest on £550 invested for 5 years at 8% per year.

2 £700 is invested for 3 years and is then worth £836·50. What was the rate of simple interest?

3 Money in a deposit account earns simple interest at a rate of 8% per annum. If I put £400 in my deposit account, what will it be worth after one year?

4 Find the simple interest on £2600 invested for 5 years at 5% per year.

5 A businessman lends money at an interest rate of $9\frac{1}{2}$% per annum. The sum he lends is £4750. After a certain time it will earn him £1805. What is the period of time of the loan?

6 A certain sum of money is invested at 10% per year for 4 years and is then worth £2240. How much was originally invested?

7 A businessman invests money at 15% per annum for 2 years. At the end of this time the money amounts to £1950. How much money did he invest?

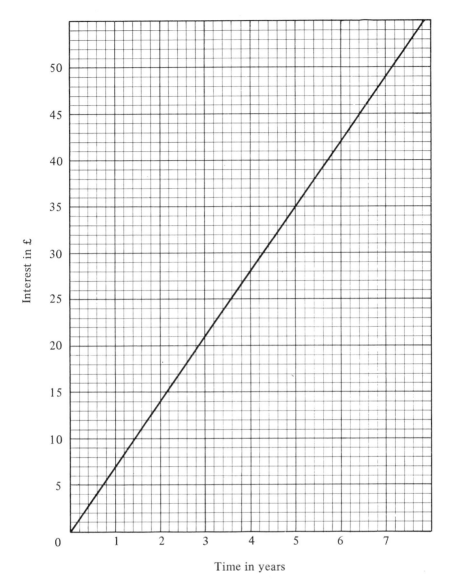

Interest in £

Time in years

8 The graph above shows the simple interest received when £100 is invested for a number of years at 7% per annum.
Use the graph to find:

(a) the simple interest received on £100 invested for 4 years at 7% per annum;

(b) the number of years £100 would have to be invested at 7% per annum to receive £42 interest;

(c) the simple interest received on £300 invested for 3 years at 7% per annum;

(d) the simple interest received on £500 invested for 5 years at $3\frac{1}{2}$% per annum.

7

Exercise 3. Compound interest

1 A man invests £1000 at a rate of 5% per annum.

(a) What interest does he earn in one year?
(b) What is his total investment after this first year?
(c) He leaves all the money invested and earns interest on this total sum after a further year. How much interest has he accumulated at the end of the second year?
(d) How much money does he now have altogether?
(e) If he leaves the total sum invested for another year (i.e. a third year) how much interest will he have earned at the end of this third year? (Answer to nearest penny.)

2 A man puts £100 in the bank on 1 January 1977. He leaves it there for 3 complete years, drawing it out on the last day of 1979. The bank pays 4% per annum compound interest on the money. How much will he draw out altogether? (Answer to the nearest penny.)

3 Find the compound interest on £6000 for 2 years at 8% per annum.

4 A businessman invested £1600 for 3 years at 9% per annum compound interest. What was the investment then worth?

5 A businessman invested £1600 for 3 years at 8% per annum compound interest. What was the investment then worth?

6 A house is purchased for £22 000 in 1977. The value of the house increases at a rate of 10% per year. What will the value of the house be 3 years later in 1980?

7 A house is insured in 1977 at a premium of £30. The premium is increased by 8% each year. What will the premium be 3 years later in 1980? (Answer to the nearest penny.)

8 A large business borrows £500 000 to complete a shopping centre. The rate of interest is 5% per annum and the loan is to be repaid after 3 years.

(a) If the interest is to be calculated as simple interest how much interest would the business have to pay?
(b) If the interest is to be calculated as compound interest how much interest would the business have to pay?
(c) What is the difference between these payments?
(d) What rate of simple interest would produce the same amount of interest as the compound interest did? (Give your answer correct to the third decimal place.)

9

Years	10% per annum	11% per annum	12% per annum	13% per annum	14% per annum
1	£1·100 00	£1·110 00	£1·120 00	£1·130 00	£1·140 00
2	£1·210 00	£1·232 10	£1·254 40	£1·276 90	£1·299 60
3	£1·331 00	£1·367 63	£1·404 93	£1·442 90	£1·481 54
4	£1·464 10	£1·518 07	£1·573 52	£1·630 48	£1·688 96

The table above shows the amount to which £1 grows at compound interest for the time and rates given. Use this information to answer the following questions.

(a) What interest, in £, is obtained when £1000 is invested for 3 years at 12% per annum?

(b) What interest, in £, is obtained when £12 000 is invested for 4 years at 14% per annum compound interest?

(c) What sum, invested for 4 years at 11% per annum, amounts to £15 180·7?

(d) What sum, invested for 3 years at 10% per annum, amounts to £2662?

(e) What sum, invested for 2 years at 14% per annum, amounts to £38 988?

(f) When £500 is invested for 1 year at a certain rate of compound interest it amounts to £565. What rate of compound interest is being paid?

(g) What sum, invested for 2 years at 12% per annum, yields £101·76 interest?

(h) What sum, invested for 4 years at 11% per annum, yields £3108·42 interest?

2 Borrowing Money

Exercise 4. Banks

Amount of Loan	6 Months		12 Months		18 Months		24 Months		Amount of Loan
	Interest	Monthly repayment	Interest	Monthly repayment	Interest	Monthly repayment	Interest	Monthly repayment	
£	£	£	£	£	£	£	£	£	£
100	5.00	17.50	10.00	9.17	15.00	6.39	20.00	5.00	100
200	10.00	35.00	20.00	18.34	30.00	12.78	40.00	10.00	200
300	15.00	52.50	30.00	27.50	45.00	19.17	60.00	15.00	300
400	20.00	70.00	40.00	36.67	60.00	25.56	80.00	20.00	400
500	25.00	87.50	50.00	45.84	75.00	31.95	100.00	25.00	500
600	30.00	105.00	60.00	55.00	90.00	38.34	120.00	30.00	600
700	35.00	122.50	70.00	64.17	105.00	44.73	140.00	35.00	700
800	40.00	140.00	80.00	73.34	120.00	51.12	160.00	40.00	800
900	45.00	157.50	'90.00	82.50	135.00	57.50	180.00	45.00	900
1000	50.00	175.00	100.00	91.67	150.00	63.89	200.00	50.00	1000
1100	55.00	192.50	110.00	100.84	165.00	70.28	220.00	55.00	1100
1200	60.00	210.00	120.00	110.00	180.00	76.67	240.00	60.00	1200
1300	65.00	227.50	130.00	119.17	195.00	83.06	260.00	65.00	1300
1400	70.00	245.00	140.00	128.34	210.00	89.45	280.00	70.00	1400
1500	75.00	262.50	150.00	137.50	225.00	95.84	300.00	75.00	1500

Amount of Loan	30 Months		36 Months		48 Months		60 Months		Amount of Loan
	Interest	Monthly repayment	Interest	Monthly repayment	Interest	Monthly repayment	Interest	Monthly repayment	
£	£	£	£	£	£	£	£	£	£
100	–	–	–	–	–	–	–	–	100
200	50.00	8.34	60.00	7.23	80.00	5.84	100.00	5.00	200
300	75.00	12.50	90.00	10.84	120.00	8.75	150.00	7.50	300
400	100.00	16.67	120.00	14.45	160.00	11.67	200.00	10.00	400
500	125.00	20.84	150.00	18.06	200.00	14.59	250.00	12.50	500
600	150.00	25.00	180.00	21.67	240.00	17.50	300.00	15.00	600
700	175.00	29.17	210.00	25.28	280.00	20.42	350.00	17.50	700
800	200.00	33.34	240.00	28.89	320.00	23.34	400.00	20.00	800
900	225.00	37.50	270.00	32.50	360.00	26.25	450.00	22.50	900
1000	250.00	41.67	300.00	36.12	400.00	29.17	500.00	25.00	1000
1100	275.00	45.84	330.00	39.73	440.00	32.09	550.00	27.50	1100
1200	300.00	50.00	360.00	43.34	480.00	35.00	600.00	30.00	1200
1300	325.00	54.17	390.00	46.95	520.00	37.92	650.00	32.50	1300
1400	350.00	58.34	420.00	50.56	560.00	40.84	700.00	35.00	1400
1500	375.00	62.50	450.00	54.17	600.00	43.75	750.00	37.50	1500

The table above shows the interest on loans over given periods and the amount the client will have to repay each month to clear the loan. The interest is based on a flat rate of 10% per annum. Use the table to answer the following questions.

1 A businessman borrows £1000. What interest will he pay if he takes the loan over the following times?

(*a*) 12 months   (*b*) 24 months   (*c*) 30 months   (*d*) 60 months

2 What is total repayment on each of the following loans?

(*a*) £900 for 30 months   (*b*) £1300 for 18 months
(*c*) £700 for 24 months   (*d*) £1200 for 60 months

10

3 A businessman borrows £1400. What interest will he pay and what will his monthly repayments be if he takes the loan over the following times?

(a) 18 months   (b) 30 months   (c) 48 months   (d) 60 months

4 A client wants to borrow £1300 but wants to pay back less than £70 each month. What is the least time he can take the loan over?

5 Find the least time for the loans with maximum monthly payments listed below.

	Loan	Monthly payment *less than*
a	£1200	£40
b	£800	£35
c	£1500	£40
d	£900	£100
e	£1100	£90

6 In order to replace old machinery, a farmer borrows £4500 at a flat rate of 10% per annum over a period of three years.

(a) What interest does he pay on the loan?
(b) What is the amount of the total repayment?
(c) What would his monthly repayments be?

Exercise 5. Building Societies and mortgages

1 Building Societies pay interest on money deposited with them at stated rates of interest. The income tax on the interest is paid by the society. In the following examples, calculate how much interest a depositor will earn after one year.

	Amount deposited	Interest rate (*Tax paid*)		Amount deposited	Interest rate (*Tax paid*)
a	£20 000	9% per year	f	£900	8·6% per year
b	£1 600	10% per year	g	£8 500	9·4% per year
c	£1 300	8·5% per year	h	£14 000	8·25% per year
d	£2 750	8·75% per year	i	£3 200	9% per year
e	£5 200	9·25% per year	j	£1 800	8·3% per year

2 A Building Society advertises that its rate of interest is equivalent to 13% before tax is paid. A client deposits £1000.

(a) Calculate the interest payable on this sum after one year.
(b) Calculate how much tax has to be paid on this at a rate of 30%.
(c) How much interest will the client actually receive?

3 Follow the method of Question 2 to work out how much a client will actually receive in interest at the end of one year in the following examples.

Amount deposited	Rate of interest (*tax to be paid*)	Amount deposited	Rate of interest (*tax to be paid*)
a	£1 500	13% per year	f	£18 500	13·2% per year
b	£1 500	14% per year	g	£14 000	12·9% per year
c	£1 500	15% per year	h	£950	14·2% per year
d	£3 000	12·5% per year	i	£3 400	11·9% per year
e	£7 500	13·4% per year	j	£5 700	13·8% per year

4 A butcher wishes to buy a new shop costing £33 000. The Building Society agrees to lend him 80% of this sum at the rate of 104p per £100 per month over 25 years.

(*a*) What is the amount of the deposit the butcher has to pay?
(*b*) How much is the loan?
(*c*) What is the monthly payment for the borrower?
(*d*) What is the total eventual cost of the loan?

5 A businessman decides to invest in property. He buys four flats costing £9500, £12 400, £13 700, £19 800. He borrows money from the Building Society to do so. The period of the loans is 20 years and the rate is 102p per £100 (or *pro rata* for part of £100) per month. He is required to put down an initial deposit of 30%. *Taking each flat separately*, calculate

(*a*) the amount borrowed,
(*b*) the interest paid per month.

Exercise 6. Hire purchase

1 In the examples below, find the *monthly* instalments the purchaser has to pay. Round your answers up to a whole penny.

	Selling price of goods	Deposit	Rate of interest	Number of years
a	£35	£7	15%	1
b	£27	£7	10%	1
c	£340	£40	15%	2
d	£110	£30	12·5%	2
e	£185	£25	17·5%	2
f	£63	£13	8%	1
g	£549	£129	20%	2
h	£111	£20	12·5%	2
i	£658	£10	8%	1
j	£43	£7	15%	1

2 In the examples below, find the *weekly* instalments the purchaser has to pay. Round your answers up to a whole penny.

	Selling price of goods	Deposit	Rate of interest	Number of years
a	£140	£20	15%	1
b	£180	£40	17·5%	1
c	£199	£39	12·5%	1
d	£329	£49	12·5%	1
e	£449	£99	22·5%	2
f	£640	£100	15%	2
g	£999	£99	10%	2
h	£1250	£350	8%	2
i	£1300	£200	26%	1
j	£1450	£400	28%	1

3 The price of a washing machine is £344. A shop offers a discount of 15% for cash. If a customer decided to buy by hire purchase he is asked to make a deposit of 10% and pay the rest by monthly instalments. These instalments include an interest charge of 12·5% per year.

(*a*) What will a customer pay if he pays cash?
(*b*) If he pays by hire purchase, how much is his deposit?
(*c*) How much interest does he pay on the remaining value if he elects to pay it off over 2 years?
(*d*) What is the amount of his monthly instalments?
(*e*) What is the difference in money between paying cash and buying by hire purchase?

4 The cash price of a portable TV set was £84·50. The shop asks for a deposit of £30, 8 monthly payments of £6·50 and a final payment of £7.

(*a*) What is the total hire purchase cost?
(*b*) What is the difference between the HP and the cash price?
(*c*) Calculate this difference as a percentage of the amount loaned to the purchaser.
(*d*) What flat rate of interest *per year* is the shop charging?

5 A colour television is priced at £399. If bought for cash, a discount of 20% is given. The hire purchase (HP) terms are a deposit of £60 and 12 equal monthly payments of £31·64. Find:

(*a*) the cash price of the television,
(*b*) the total hire purchase (HP) price of the television.
The same shop rents a similarly priced television for £11·25 a month.
(*c*) How much will the set earn for the shop in a year?
(*d*) How many years will the set have to be rented to recover the full price of £399?

THREE EASY WAYS to a

SUPER PANASONIC VIDEO CASSETTE RECORDER

1. Buy one for only £699 cash price

2. Buy one with 9 monthly payments of £83.00

3. Rent one for £113.46 deposit (6 months advance rental) and £18.91 per month thereafter

Established 1900

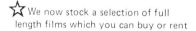 We now stock a selection of full length films which you can buy or rent

6 (*a*) What is the cost of the VCR (video cassette recorder) if 9 monthly payments are made?
   (*b*) What is the difference between the cash price and the HP price?
   (*c*) What rate of interest *per year* is being charged? (Remember the loan is for 9 months only, i.e. $\frac{3}{4}$ of a year.)
   (*d*) What is the cost of renting a VCR for 1 year?

Exercise 7. Depreciation

1 A factory re-equips with new machinery at the start of the financial year. The value of a machine falls each year of its life. Calculate the value of each machine after one year for the depreciation shown.

	Cost of machine	Depreciation		Cost of machine	Depreciation
a	£12 000	20%	f	£16 000	8%
b	£48 000	25%	g	£19 000	22%
c	£36 000	18%	h	£64 000	16%
d	£80 000	15%	i	£17 000	9%
e	£10 000	$12\frac{1}{2}$%	j	£21 000	14%

2 A factory 'writes off' the value of its machinery over a period of years. Each year the machinery is calculated to have depreciated by a fixed percentage. Calculate the value of each machine after the stated number of years in the examples given in the table. Work to the nearest £1 at each stage where necessary.

	Cost of machine	Fixed depreciation	Number of years
a	£12 000	20%	2
b	£48 000	25%	2
c	£20 000	15%	3
d	£9 000	$12\frac{1}{2}$%	3
e	£30 000	8%	2
f	£24 000	6%	2
g	£10 000	7%	3
h	£80 000	9%	3
i	£66 000	12%	3
j	£42 000	14%	2

3 A car leasing firm has three ranges of car for hire. Each car in its first year depreciates in value by 20%. In the second year it depreciates by 15% of its value at the beginning of the year, and in the third year its depreciation is 10% of its value at the beginning of that year.

Range	Large car	Middle car	Small car
Cost when new	£6200	£4600	£2800

For each car, answer the following questions.

(a) What value does the car lose in its first year?
(b) What value does the car lose in its second year?
(c) What is the value of the car at the beginning of the fourth year?

4 A shop rents out colour television sets. The capital cost of each one new is £380. The set depreciates by 25% in its first year. In its second year it depreciates by 20% of the value it had at the end of the first year. Find the value of the television set

(*a*) at the end of the first year,
(*b*) at the end of the second year.

5 The value of a certain machine when new was £14 000. In the first year it depreciated by 12% of its value when new. In the second year it depreciated by 5% of its value at the beginning of the second year.

Express the value of this machine after *two* years as a percentage of its original value.

Exercise 8. Credit cards

1 A credit card scheme adds interest each *month* at a rate of $1\frac{1}{2}\%$ of the balance outstanding.

(*a*) What is the interest on £100 for 1 month?
(*b*) Copy and complete the following table which shows the amount the creditor owes on the 1st June and the amount he pays back over three months.

Balance on 1 June	. . . . .	£100·00
Interest added for June	. . . . .	
Repayment on 1 July	. . . . .	£41·50
Balance after repayment	. . . . .	
Interest added for July	. . . . .	
Repayment on 1 August	. . . . .	£20·90
Balance after repayment	. . . . .	
Interest added for August	. . . . .	
Repayment on 1 September	. . . . .	£20·60
Balance after repayment	. . . . .	
Interest added for September	. . . . .	

(*c*) The creditor repaid the whole of the balance outstanding at the end of September.

(i) How much did the man pay altogether?
(ii) How much interest did he pay?
(iii) What percentage of his initial loan is this interest?

(*d*) One of the rules for the use of credit cards states that the smallest amount which the creditor can repay is 15% of the balance outstanding or £6·00 whichever is the greater. Calculate the largest balance that can be outstanding if the smallest amount is to be repaid.

2 Because of changes introduced in the Budget a credit card company raises its monthly interest charge to 2·00% (i.e. 2% per month on any balance).

(a) Copy and complete the following table.

Balance on 1 January	. . . . .	£450·00
Interest added for January	. . . . .	
Repayment on 1 February	. . . . .	£59·00
Balance after repayment	. . . . .	
Interest added for February	. . . . .	
Repayment on 1 March	. . . . .	£88.00
Balance after repayment	. . . . .	
Interest added for March	. . . . .	
Repayment on 1 April	. . . . .	£176·40
Balance after repayment	. . . . .	
Interest added for April	. . . . .	

(b) How much does the creditor owe now?
(c) How much interest has he paid so far?

3 Buying and selling

Exercise 9. Profit and loss, discounts

1 The cost prices of various goods are given below. The profit on each one is to be 25%. Work out the selling prices.

(a) 32p   (b) 96p   (c) £1·74   (d) £6·82   (e) £74·90
(f) £124·60   (g) £234·52   (h) £409·92   (i) £642·20   (j) £174·40

2 The cost prices of various goods are given below. The profit on each one is to be 120%. Work out the selling prices.

(a) 80p   (b) £2·20   (c) £3·85   (d) £9·90   (e) £12·35
(f) £18·40   (g) £27·70   (h) £127·70   (i) £345·50   (j) £2340

In the retail trade many calculations are based on the selling price rather than the cost price. Thus if an article sells at £1·20 and the *profit margin* is $33\frac{1}{3}$%, the cost price will be 80p.

3 The selling prices of various goods are given below. The profit margin on each one is $33\frac{1}{3}$%. Work out the cost prices.

(a) 90p   (b) £1·80   (c) £2·37   (d) £5·07   (e) £9·99
(f) £15·15   (g) £23·13   (h) £38·58   (i) £174·99   (j) £1248

4 The selling prices of various goods are given below. The profit margin on each on is to be 55%. Work out the cost prices to the nearest penny.

(a) £1·80   (b) £2·78   (c) £3·40   (d) £6·60   (e) £8·44
(f) £11·99   (g) £39·39   (h) £120   (i) £240   (j) £1800

5 A retailer gives a cash discount to customers who pay cash straight away. Some list prices and cash discounts are given below. In each example, calculate to the nearest penny, how much the customer actually pays.

	List price	Cash discount		List price	Cash discount
a	£10·80	10%	e	£8·90	12%
b	£12·70	15%	f	£27·50	14%
c	£16·40	8%	g	£38·99	10%
d	£1·20	25%	h	£124	$12\frac{1}{2}$%

6 A retailer gets a trade discount when he buys his goods from the distributors so he pays less than the stated price. In each example below, calculate, to the nearest penny, how much the retailer actually pays.

	Price	Trade discount		Price	Trade discount
a	£75	20%	e	£445	16%
b	£300	$33\frac{1}{3}$%	f	£84	$12\frac{1}{2}$%
c	£450	20%	g	£270	15%
d	£212	25%	h	£850	21%

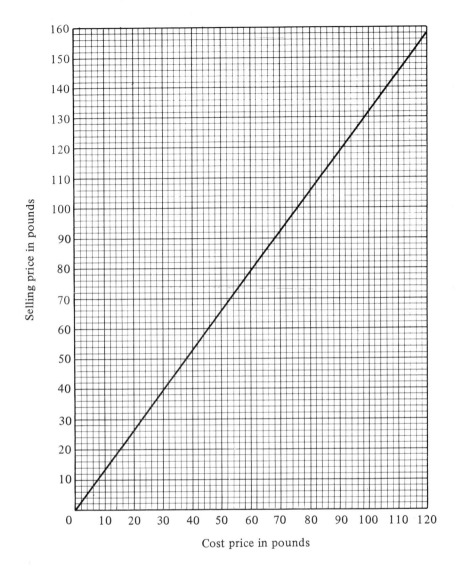

Selling price in pounds

Cost price in pounds

The graph shows the selling price of articles giving a profit of 32% on the cost price.

7 Use the graph to find the selling prices that correspond to the following cost prices.

(*a*) £20   (*b*) £50   (*c*) £100   (*d*) £14   (*e*) £36   (*f*) £108   (*g*) £92
(*h*) £25   (*i*) £63   (*j*) £112

8 Use the graph to find the cost prices that correspond to the following selling prices.

(*a*) £50   (*b*) £66   (*c*) £132   (*d*) £25   (*e*) £40   (*f*) £120   (*g*) £34
(*h*) £72   (*i*) £87   (*j*) £143

Exercise 10. Percentage profits and discounts

1 A trader buys music centres at £228·65. He sells them at £269. What is his profit? What percentage profit on the cost price is this?

2 A trader buys automatic washing machines at £159·96 and sells them at £199·95. What is his profit? What percentage profit on the cost price is this?

3 A trader buys video recorders at £645 and sells them at £750. What is his profit? What percentage profit on the cost price is this?

4 Fridge/freezers are sold to a customer who puts down a deposit of £35·96 and makes 11 monthly payments of £13·09. How much does the customer pay?

The trader bought the fridge/freezers at £135. What is his profit? What percentage profit on the cost price is this to the nearest whole per cent?

5 To buy a small colour TV set, a customer puts down a deposit of £59·95 and then makes 11 monthly payments of £21·81. How much does he pay altogether?

The shop bought the TV sets at a cost of £210. What is its profit? What percentage profit on the cost price is this to the nearest per cent?

6 A customer puts down a deposit of £35·96 on a twin-tub washing machines and then makes 11 monthly payments of £13·09. How much does he pay altogether?

The shop bought the washing machines for £117. What is its profit? What percentage profit on the cost price is this to the nearest per cent?

7 A retailer buys freezers at a list price of £279·95 less a trade discount of $33\frac{1}{3}\%$. What do they cost him to the nearest penny? What is his profit if he sells them at the list price? What percentage profit on his cost price is this?

8 A retailer buys cameras at a list price of £52·65 less a trade discount of 25%. What does each camera cost him to the nearest penny? If he sells at the list price what is his profit? He decides to sell them at a sale price of £45. What is his percentage profit on this selling price to the second decimal place?

9 A set of golf clubs is listed by the distributors at a price of £181·45. A retailer purchases them at a trade discount of 30%. What does the set cost him to the nearest £1? If he sells the set at the list price, what is his profit? In fact he sells them at a cash discount of 10% on the list price. What is his selling price? What is his percentage profit on his selling price to the second decimal place?

10 A businessman calculates that in the next financial year his total production costs will be £140 000. He also calculates he will produce 50 000 articles for sale. What price should he charge per article if he wishes to make a 10% profit?

11 A businessman calculates that in the next financial year his total production costs will be £276 000. His firm will produce 40 000 articles for these costs. He wishes to make a profit of 20%. What price should be charged for each article?

12 A businessman makes a forecast of his profits for the next business year by listing the following information.

Cost of raw materials	£300 000 ± 10%
Labour costs	£150 000 ± 15%
Other costs	£80 000 ± 10%
Number of articles produced	90 000 ± 8%
Selling price per article	£8 ± 50p

(*a*) What are the possible maximum total costs?
(*b*) What are the possible minimum total costs?
(*c*) What is the greatest number of articles he might produce?
(*d*) What is the minimum number of articles he might produce?
(*e*) How much money will he earn if he produces the greatest number of articles and sells them at the maximum selling price?
(*f*) How much money will he earn if he produces the least number of articles and sells them at the minimum selling price?
(*g*) What is his greatest possible profit?
(*h*) What is his least possible profit?

13 The following year the businessman makes a further forecast by listing the information below. Costs in general have gone up but a stricter control is to be maintained on the production time.

Cost of raw materials	£350 000 ± 10%
Labour costs	£180 000 ± 15%
Other costs	£100 000 ± 10%
Number of articles produced	110 000 ± 5%
Selling price per article	£10 ± 50p

(*a*) What is his greatest possible profit?
(*b*) What is his least possible profit?

14 A shopkeeper adds 25% to the cost price of an article to give him his profit. He then adds 15% to this new price to allow for VAT (value added tax) which gives him the normal selling price. In a sale he takes 10% off the selling price of all goods. Copy the following table and complete it. (Work to the nearest penny where appropriate.)

	Cost price	Normal selling price	Sale price
a	£4·20		
b		£6·60	
c			£8·50
d	£10·25		
e		£14·75	
f			£19·99

15 Another shopkeeper adds 30% to the cost price, and then 15% for VAT on top of that to get his normal selling price. In a sale he deducts 8% of this normal selling price. Copy the following table and complete it.

	Cost price	Normal selling price	Sale price
a	£2·30		
b		£4·64	
c			£6·99
d	£11·43		
e		£15·63	
f			£20·20

16 A shopkeeper adds 35% to the cost price of goods, and then 15% for VAT on top of that to get his normal selling price. In a sale he reduces his normal selling price by 20%. Copy the following table and complete it.

	Cost price	Normal selling price	Sale price
a	£3·50		
b		£9·60	
c			£12·40
d	£2·25		
e		£8·90	
f			£24·50

17 A firm employs 140 men and women who earn an average wage of £64 per week. Wage costs are 40% of the total costs of production. The firm produces 180 washing machines per week. The charge to the retailer is £164·50 per machine.

(*a*) What are the total production costs per week of the firm?

(*b*) What is the total profit the firm makes in one week?

(*c*) What is the profit as a percentage of the production costs to the nearest whole number?

(*d*) The retailer adds 35% to his cost price to establish the selling price. What is the selling price?

(*e*) The retailer must add 15% to the selling price for VAT. What does the customer have to pay?

18 A firm employs 220 men and women who earn an average weekly wage of £73. Wage costs are 60% of the total costs of production. The firm produces 300 articles per week. The cost to the retailer is £164·30 per article.

(*a*) What is the firm's profit as a percentage of the production costs to the nearest whole number?

(*b*) The retailer adds 30% to his cost price to establish his profit and a further 15% for VAT. What is the selling price?

(*c*) At a sale, the retailer reduces the price of each article by 25%. What is the difference between this new price and the cost price?

19 A retailer buys goods from a manufacturer who allows him a trade discount of 25% off the list price which is £16·40 per article. A customer when buying one of these items from the retailer is allowed a cash discount of 5p in the pound off the list price. What is the actual profit of the retailer? What percentage profit is this of the cost price to the nearest whole number?

20 A retailer buys goods from a manufacturer who allows him a trade discount of 30% of the list price which is £145·90. What does the retailer pay? He allows a customer a cash discount of 20% off the list price. What does the customer pay? What is the retailer's actual profit? What percentage profit is this of the cost price to the retailer to the nearest whole number?

4 Earnings, taxes, rates

Exercise 11. Payment of wages, rates of pay

1 A firm employs staff who may have different 'normal working hours' and who also work different hours of overtime. The rates of pay vary for individual members of staff. Overtime is paid at time and a half. Calculate the gross wage for each member of staff from the information in the table below.

	Normal hours worked	Hours of overtime worked	Rate per hour £
a	40	10	1·10
b	40	8	1·10
c	40	12	1·10
d	40	7	1·10
e	40	5	1·10
f	35	10	1·25
g	35	5	1·25
h	35	14	1·25
i	35	9	1·25
j	35	6	1·25
k	33	7	1·35
l	33	8	1·35
m	33	14	1·35
n	33	9	1·35
o	33	5	1·35
p	38	12	1·75
q	38	14	1·75
r	38	16	1·75
s	38	2	1·75
t	38	3	2·20
u	42	7	2·20
v	42	12	2·20
w	42	6	2·20
x	42	5	2·20
y	42	4	2·20
z	32	6	0·94
A	32	7	0·94
B	32	10	0·94
C	32	16	0·94
D	32	5	0·94

1	0·87	41	35·67	81	70·47	121	105·27	161	140·07
2	1·74	42	36·54	82	71·34	122	106·14	162	140·94
3	2·61	43	37·41	83	72·21	123	107·01	163	141·81
4	3·48	44	38·28	84	73·08	124	107·88	164	142·68
5	4·35	45	39·15	85	73·95	125	108·75	165	143·55
6	5·22	46	40·02	86	74·82	126	109·62	166	144·42
7	6·09	47	40·89	87	75·69	127	110·44	167	145·29
8	6·96	48	41·76	88	76·56	128	111·36	168	146·16
9	7·83	49	42·63	89	77·43	129	112·23	169	147·03
10	8·70	50	43·50	90	78·30	130	113·10	170	147·90
11	9·57	51	44·37	91	79·17	131	113·97	171	148·77
12	10·44	52	45·24	92	80·04	132	114·84	172	149·64
13	11·31	53	46·11	93	80·91	133	115·71	173	150·51
14	12·18	54	46·98	94	81·78	134	116·58	174	151·38
15	13·05	55	47·85	95	82·65	135	117·45	175	152·25
16	13·92	56	48·72	96	83·52	136	118·32	176	153·12
17	14·79	57	49·59	97	84·39	137	119·19	177	153·99
18	15·66	58	50·46	98	85·26	138	120·06	178	154·86
19	16·53	59	51·33	99	86·13	139	120·93	179	155·73
20	17·40	60	52·20	100	87·00	140	121·80	180	156·60
21	18·27	61	53·07	101	87·87	141	122·67	181	157·47
22	19·14	62	53·94	102	88·74	142	123·54	182	158·34
23	20·01	63	54·81	103	84·61	143	124·41	183	159·21
24	20·88	64	53·68	104	90·48	144	125·28	184	160·08
25	21·75	65	56·55	105	91·35	145	126·15	185	160·95
26	22·62	66	57·42	106	92·22	146	127·02	186	161·82
27	23·49	67	58·29	107	93·09	147	127·89	187	162·69
28	24·36	68	59·16	108	93·96	148	128·76	188	163·56
29	25·23	69	60·03	109	94·83	149	129·63	189	164·43
30	26·10	70	60·90	110	95·70	150	130·50	190	165·30
31	26·97	71	61·77	111	96·57	151	131·37	191	166·17
32	27·84	72	62·64	112	97·54	152	132·24	192	167·04
33	28·71	73	63·51	113	98·31	153	133·11	193	167·91
34	29·58	74	64·38	114	99·18	154	133·98	194	168·78
35	30·45	75	65·25	115	100·05	155	134·85	195	169·65
36	31·32	76	66·12	116	100·92	156	135·72	196	170·52
37	32·19	77	66·99	117	101·79	157	136·59	197	171·39
38	33·06	78	67·86	118	102·66	158	137·46	198	172·26
39	33·93	79	68·73	119	103·53	159	138·33	199	133·13
40	34·80	80	69·60	120	104·40	160	139·20	200	174·00

2 A firm pays its apprentices 87p an hour. Use the ready reckoner on page 25 to answer the following questions.

(a) How much will an apprentice earn in an 8-hour day?
(b) How much will he earn in a 5-day week of 8 hours per day?
(c) The firm has 17 apprentices. How much is its daily wage bill for them?
(d) One apprentice works the equivalent of a 57-hour week at the standard rate per hour. How much does he earn?
(e) Three apprentices between them work 189 hours at the standard rate. What are their total earnings? If these earnings are equally divided between them how much does each get?

3 A journeyman earns £2·87 per hour. Use the ready reckoner on page 25 to answer the following questions.

(a) How much does he earn in a 7-hour day?
(b) How much does he earn in a 35-hour week?
(c) A particular job involves 4 journeymen who work a total of 14 hours. What is the wage bill for the job?

4 The following sums of money show how much an apprentice was paid for a number of different jobs at the standard rate of 87p per hour. Use the ready reckoner to find how many hours he worked on each job.

(a) £5·22   (b) £160·08   (c) £84·39   (d) £7·83   (e) £26·97

5 An apprentice is paid time and half for work on Saturday and double time for Sunday work. Use the ready reckoner to calculate how much he earns in the weeks shown below.

(a) Normal working – 37 hours, Saturday – 3 hours, Sunday – 3 hours
(b) Normal working – 37 hours, Saturday – 2 hours, Sunday – 6 hours
(c) Normal working – 35 hours, Saturday – 6 hours, Sunday – 5 hours
(d) Normal working – 38 hours, Saturday – 3 hours, Sunday – 3 hours
(e) Normal working – 36 hours, Saturday – 5 hours

6 A factory pays a particular group of its workers £1·34 per hour. The agreed working week is 36 hours. However, every worker is expected to do 6 hours overtime per week for which he is paid at time and a half. There are 70 workers involved. What is the total salary bill for these workers?

7 A factory pays its assembly-line workers £1·63 per hour. One worker works his normal week of 38 hours. He works 8 hours overtime for which he is paid at time and a half. He also works 6 hours on Sunday for which he is paid double time.

(a) What is his wage for the week?
(b) If this worker's wage is taken as the average wage, what would the factory's wage bill be if there are 56 workers on the assembly line?

8 A lorry-driver is paid £1·60 per hour for a 38-hour week. Any driving done on Sunday is paid at double time. Overtime on any other day is paid at time and a quarter.

(a) What is his weekly wage if he works no overtime?
(b) How much extra does he earn if he works 4 hours overtime on Monday?
(c) How much does he earn in a week if he works 3 hours overtime on Wednesday, 2 hours overtime on Friday and 6 hours overtime on Sunday?

9 A factory worker is paid 127p per hour for a basic 40-hour week. She is paid time and a half for overtime worked from Monday to Saturday, and double time on Sunday.

(a) What is her basic weekly wage when she does not work any overtime?
(b) What is her gross wage for working 49 hours in a particular week, made up of 43 hours from Monday to Saturday, together with 7 hours on Sunday?
(c) In the following week, during which she worked 40 hours from Monday to Saturday, her gross wage was £63·50. How many hours did she work on Sunday?
(d) The next week her gross wage if £73·66. She did no overtime on Sunday. How many hours of overtime did she work during the week?

10 A factory makes each worker clock in and clock out each day. This is recorded as shown below.

	In	Out	In	Out
Monday	7.20 a.m.	12.31 p.m.	1.28 p.m.	4.31 p.m.
Tuesday	7.25 a.m.	12.32 p.m.	1.29 p.m.	4.31 p.m.
Wednesday	7.20 a.m.	12.18 p.m.	1.50 p.m.	4.36 p.m.
Thursday	7.35 a.m.	12.35 p.m.	1.31 p.m.	7.32 p.m.
Friday	7.40 a.m.	12.31 p.m.	1.28 p.m.	4.31 p.m.
Saturday	8.57 a.m.	1.01 p.m.		
Sunday	8.50 a.m.	1.19 p.m.		

Each worker is paid from the nearest quarter of an hour after his arrival to the nearest quarter of an hour before departure. (For example, if he arrives at 7.35 a.m. he will not start being paid until 7.45 a.m. and if he leaves at 4.40 p.m. he will only paid to 4.30 p.m.)

The basic rate of pay is £2·20 per hour. This rate is paid for hours worked up to 8 hours on each day from Monday to Friday. Any time worked more than that on these days is paid at an overtime rate of time and a half. Overtime worked on Saturday is paid at time and three-quarters and on Sunday at double time.

(a) Copy and complete the table below.

	Basic hours worked	Hours of overtime worked
Monday		
Tuesday		
Wednesday		
Thursday		
Friday		
Saturday		
Sunday		

(b) What is the gross pay the worker earns this week?
(c) He has deductions from this gross pay of £26·40. What does he receive in his pay packet?
(d) If this week is an average working week, what will the worker earn as his gross pay and as his net pay in one year of 52 weeks?
(e) What do the deductions amount to in one year?

Exercise 12. Commission

1 A department store gives its assistants a basic weekly wage. They are expected to sell a certain value of goods each week (their quota). If they sell more than this value then they earn commission on the extra sales. From the table below, calculate weekly wages.

	Weekly wage	Quota	Actual sales	Commission
a	£30·20	£200	£250	6%
b	35·40	300	380	6%
c	28·80	180	300	8%
d	32·60	320	360	5%
e	34·80	300	345	5%
f	38·80	350	480	4%
g	40·20	400	430	4%
h	36·60	340	420	5%
i	44·30	400	450	3%
j	46·20	450	400	2%

2 A salesman receives a basic annual salary of £1 800. He receives a 5% commission on his sales throughout the year. In one year, he sold £44 000 worth of goods. How much did he earn that year? If his commission is spread evenly throughout the year what was his monthly salary?

28

3 A salesman receives a basic annual salary of £3200. He receives a 10% commission on his sales throughout the year. In one year his sales amounted to £22 000. How much did he earn that year? If his commission is spread evenly throughout the year, what was his monthly salary?

4 A firm pays its sales representatives a basic annual salary of £2400 plus a 6% commission on all sales. Smith, in one year, sold £33 000 worth of goods, Jones sold £40 000, and Mitchell sold £42 500.

(*a*) How much did each earn that year?
(*b*) What was this calculated as a monthly salary for each salesman?
(*c*) What was the firm's total annual salary bill for these three salesmen?
(*d*) What was their average annual salary?
(*e*) If the firm had 37 salesmen, what was their approximate salary bill that year? (Use the average salary as calculated in (*d*).)

5 A salesman has a basic annual salary of £2700. He earns a 5% commission on all his sales. If his total salary (including commission) is three times his basic salary what is the value of his sales?

6 A salesman has a basic salary of £2800. His salary with commission is £4900. His sales for the year amount to £30 000. What percentage commission is he being allowed on his sales?

7 A salesman has a basic salary of £2100. His salary with commission is £5760. His sales for the year amount to £61 000. What percentage commission is he being allowed on sales?

Exercise 13. Local finance

1 A city levies a rate of 67p in the pound. What are the rates payable on the following buildings?

(*a*) House rated at £300   (*b*) House rated at £497
(*c*) Shop rated at £640   (*d*) Garage rated at £11 340
(*e*) Factory rated at £56 900   (*f*) Department store rated at £114 000

2 The rateable values of various districts are given below. Write down how much is raised on each for a rate of one penny in the pound.

(*a*) £240 000   (*b*) £380 000   (*c*) £1 400 000
(*d*) £36 343 000   (*e*) £81 568 000   (*f*) £110 347 432

3 Calculate the rate per pound that must be levied by the following districts to raise the money required to run the districts. (Work to the nearest necessary penny.)

District	a	b	c	d	e
Rateable value	£300 000	£1 240 000	£3 560 000	£17 286 000	£50 347 000
Sum to be raised	£159 000	£607 600	£1 993 600	£11 063 040	£34 739 430

4 A large factory is valued for rating purposes at £160 432. However, because of industrial de-rating, this figure is halved. The rate in the district is 72p in the £. What are the rates the factory has to pay? The following year the rate per £ goes up by 3p. What rates does the factory now pay? What is the percentage increase in the amount paid to the nearest whole number?

5 The money spent by a council is split up as shown below.

Education	56p
Police	4p
Roads	10p
Social work	9p
Lighting	2p
Drainage	8p
Other services	34p
Total	123p

This money is obtained from a general rate of 67p in the £ and a government grant of 56p in the £.
The total rateable value of the council's area is £109 815 000.

(a) How much money is raised by the general rate?
(b) How much money has been set aside to provide for social work?
(c) A man owns a house of rateable value £343. What does he pay in rates?
(d) If a householder pays £182·91 in rates, what is the rateable value of his house?

6 The table shows how the rate per £ is calculated for a particular district.

Rates per £ payable for 1978/79 in the district	District
regional rate	41p
district rate	14p
combined rate for subjects other than dwelling houses	55p
less Rate Support Grant (Domestic Element)	3p
combined rate for dwelling houses	52p
domestic water rate	8p

Note: Domestic rate reduction. The domestic element of the Rate Support Grant provides for a reduction of 3p in the £ in respect of dwelling houses and the appropriate adjustment is shown above.

(*a*) What is the rate per £ for dwelling houses (including domestic water rate)?

(*b*) What is the rate per £ for other buildings (excluding water rate)?

(*c*) A large shop is assessed at £54 432. (i) What rates will it have to pay? (ii) How much of this sum will go to the region and how much to the district? (iii) The shop is also charged a water rate of 4p per £. What does it pay in rates altogether?

(*d*) A factory has a rateable value of £142 344. This is halved for the purposes of the rates it pays because of industrial de-rating. It is charged separately for the water it uses by means of a special meter. (i) What rates does it pay? (ii) What sum goes to the region and what sum to the district?

(*e*) A house owner pays rates to the value of £274·20. (i) What is the rateable value of his house? (ii) How much of the rates he pays goes to the water rate?

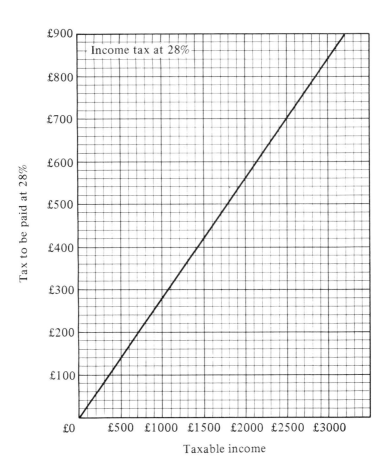

Exercise 14. Income tax

1 Use the graph to find the income tax at 28% which will be paid on the following taxable incomes. (Answers to the nearest £10.)

  (a) £1000   (b) £3000   (c) £500   (d) £2750   (e) £1750

  (f) £300   (g) £2700   (h) £1900   (i) £2400   (j) £800

2 Use the graph to find the taxable income that corresponds to the following income tax payments at 28%. (Answers to the nearest £10.)

  (a) £280 (b) £560 (c) £100 (d) £400 (e) £700 (f) £630

  (g) £210 (h) £550 (i) £690 (j) £380

3 If the rate of income tax is 30% then for every £100 of taxable income the government will take £30. Calculate how much income tax the government claims on the following taxable incomes.

  (a) £200   (b) £500   (c) £1000   (d) £1500   (e) £2700

  (f) £3400 (g) £5670 (h) £1300 (i) £3450 (j) £4590

4 If the rate of income tax is 33%, calculate how much income tax will be paid on the following taxable incomes. (Answers to the nearest pound.)

(a) £500   (b) £1000   (c) £3300   (d) £670   (e) £410
(f) £1540   (g) £5780   (h) £4700   (i) £280   (j) £355

5 Mr White receives a gross salary of £4000 a year. From this he deducts £2620 tax allowances and then he pays income tax at the rate of 30% on the remainder.

(a) Calculate (i) the amount on which he pays tax; (ii) the income tax paid; (iii) his net income after tax has been paid.
(b) He then receives an increase in gross salary of 14%. The tax allowances and the tax rate remain the same. (i) Calculate the new tax paid.
(ii) Calculate the new net income after tax has been paid.
(iii) Calculate the net increase in salary after tax has been paid.
(iv) Express this increase as a percentage of his original salary of £4000 a year.

6 A man earns £7000 per year. All his taxable pay is taxed at the standard rate of 28%. How much will he pay each month in tax if his tax allowances are as follows?

(a) Personal allowance of £1735
(b) Allowance of £912 on his Building Society interest
(c) Dependent relatives allowance of £100
(d) Expenses of £73

7 The taxation authorities investigate the tax returns of a certain man for the last financial year. Over the year the man earned £5400 and paid £610·60 tax. The allowances set against his income were as follows.

(a) Personal allowance of £1635
(b) Allowance of £1100 on Building Society interest
(c) Other allowances of £163

The taxation rate for that year was 30p in the pound on all his taxable pay. The Inland Revenue claims that since he only paid £610·60 for that year he owes them £140. Show that the Inland Revenue is justified in its claim.

8 Mr Brown earns £4900 a year. The allowances set against his salary amount to £2385. One year his taxable salary was taxed at 33%.

(a) How much tax did he pay in that year to the nearest pound?
(b) The following year the income tax dropped to 30%. How much tax did Mr Brown pay at this rate to the nearest pound?
(c) How much less did Mr Brown pay in income tax?
(d) Express this amount as a percentage of Mr Brown's total salary. Take the percentage to one decimal place.

9 A charity such as Oxfam can claim income tax back from the government or money gifted to it by individuals if certain conditions are met. In the table below, the following meanings apply.

Net income: the amount the individual gives to the charity
Tax: the amount the charity claims from the government
Gross income: the amount the individual had earned which after deduction of tax became the sum given to the charity.

Net income, tax and gross income for tax at 33%

Net Income	Tax	Gross Income	Net Income	Tax	Gross Income	Net Income	Tax	Gross Income
£	£	£	£	£	£	£	£	£
41	20·20	61·20	81	39·90	120·90	121	59·60	180·60
42	20·69	62·69	82	40·39	122·39	122	60·09	182·09
43	21·18	64·18	83	40·89	123·89	123	60·59	183·59
44	21·68	65·68	84	41·38	125·38	124	61·08	185·08
45	22·17	67·17	85	41·87	126·87	125	61·57	186·57
46	22·66	68·66	86	42·36	128·36	126	62·06	188·06
47	23·15	70·15	87	42·86	129·86	127	62·56	189·56
48	23·65	71·65	88	43·35	131·35	128	63·05	191·05
49	24·14	73·14	89	43·84	132·84	129	63·54	192·54
50	24·63	74·63	90	44·33	134·33	130	64·03	194·03
51	25·12	76·12	91	44·83	135·83	131	64·53	195·53
52	25·62	77·62	92	45·32	137·32	132	65·02	197·02
53	26·11	79·11	93	45·81	138·81	133	65·51	198·51
54	26·60	80·60	94	46·30	140·30	134	66·00	200·00
55	27·09	82·09	95	46·80	141·80	135	66·50	201·50
56	27·59	83·59	96	47·29	143·29	136	66·99	202·99
57	28·08	85·08	97	47·78	144·78	137	67·48	204·48
58	28·57	86·57	98	48·27	146·27	138	67·98	205·98
59	29·06	88·06	99	48·77	147·77	139	68·47	207·47
60	29·56	89·56	100	49·26	149·26	140	68·96	208·96
61	30·05	91·05	101	49·75	150·75	141	69·45	210·45
62	30·54	92·54	102	50·24	152·24	142	69·95	211·95
63	31·03	94·03	103	50·74	153·74	143	70·44	213·44
64	31·53	95·53	104	51·23	155·23	144	70·93	214·93
65	32·02	97·02	105	51·72	156·72	145	71·42	216·42
66	32·51	98·51	106	52·21	158·21	146	71·92	217·92
67	33·00	100·00	107	52·71	159·71	147	72·41	219·41
68	33·50	101·50	108	53·20	161·20	148	72·90	220·90
69	33·99	102·99	109	53·69	162·69	149	73·39	222·39
70	34·48	104·48	110	54·18	164·18	150	73·89	223·89
71	34·98	105·98	111	54·68	165·68	151	74·38	225·38
72	35·47	107·47	112	55·17	167·17	152	74·87	226·87
73	35·96	108·96	113	55·66	168·66	153	75·36	228·36
74	36·45	110·45	114	56·15	170·15	154	75·86	229·86
75	36·95	111·95	115	56·65	171·65	155	76·35	231·35
76	37·44	113·44	116	57·14	173·14	156	76·84	232·84
77	37·93	114·93	117	57·63	174·63	157	77·33	234·33
78	38·42	116·42	118	58·12	176·12	158	77·83	235·83
79	38·92	117·92	119	58·62	177·62	159	78·32	237·32
80	39·41	119·41	120	59·11	179·11	160	78·81	238·81

Use the table to answer the following questions.

(a) How much tax is claimed by the charity on the following gifts (net income)?

£46, £59, £100, £128, £136, £155

(b) How much money does the charity gain altogether from each gift if it claims the following amounts of tax on them?

£24·63, £33, £43·84, £77·83, £58·62, £28·57

(c) The following gifts (net income) are made to the charity. Calculate how much money the charity receives altogether.

£51, £43, £72, £110, £155, £97

Exercise 15. Value added tax (VAT)

1 The following goods are priced before VAT at 15% is added. Calculate the selling price to the nearest penny once VAT is added.

		£			£
a	Electric fire	46·50	k	Ball-point pen	3·90
b	Refrigerator	79·00	l	Fountain pen	10·45
c	Lawnmower	49·95	m	Carpet	110·30
d	Teamaker	38·95	n	Sleeping bag	13·00
e	Music centre	319·00	o	Tennis racket	14·70
f	Frying pan	3·50	p	Golf shoes	18·60
g	Vacuum cleaner	73·20	q	Jeans	12·65
h	Liquidiser	64·60	r	Pullover	7·80
i	Electric blanket	23·35	s	Calculator	18·70
j	Three-piece suite	499·95	t	Table lamp	14·20

2 The selling prices of various goods are shown below. VAT is included in the price. Calculate the price to the nearest penny before VAT is added.

		£			£
a	Lighter	6·74	k	Radio	36·40
b	Magazine	1·10	l	Fridge/freezer	212·80
c	Shirt	7·14	m	Wardrobe	156·70
d	Man's suit	74·40	n	LP record	4·65
e	Typewriter	123·90	o	Single record	1·00
f	Easy chair	246·50	p	Wrist watch	343·90
g	Executive desk	634·00	q	Bicycle	81·60
h	Rug	17·90	r	Car battery	19·20
i	Camera	84·30	s	Car tyre	17·80
j	Perfume	19·90	t	Stereo radio	110·35

3 The following bill shows the cost of a meal in a hotel. What is the total value? There is a service charge of $12\frac{1}{2}\%$. How much is the bill now? VAT at 15% is now added. What does the customer pay altogether?

$$\mathcal{C}af\acute{e}\ \mathcal{D}e\ \mathcal{L}uxe$$

Grapefruit	£0·75
Fish course	£1·32
Venison	£3·33
Vegetables	£1·80
Sweet trolley	£1·60

4 The owner of a gardening shop buys a pine seat from the wholesaler for £43. He has to pay VAT on this sum at 15%. How much more does he pay?

He then sells the seat at a price of £64 plus VAT at 15%. What is the full selling price?

The shop owner has to pay the government the difference between the amount of VAT he paid to the wholesaler and the amount collected from the customer. How much does he pay the government?

5 A boutique buys clothes from the wholesaler at a cost of £345 *including* VAT at 15%. How much VAT does the owner pay?

The clothes are then sold for a total price £517·50, again including VAT. How much is the VAT?

What does the owner pay the government?

6 The following examples give prices charged by a wholesaler and a retailer for goods. VAT has to be added to each price at 15%. In each example, calculate
(i) the VAT payable on each item,
(ii) the amount the retailer pays to the government in VAT.

	Wholesaler's price	Retailer's price		Wholesaler's price	Retailer's price
a	£3·40	£5·20	f	£343	£469
b	£7·90	£10·40	g	£112	£135
c	£18·30	£24·10	h	£99	£129
d	£27·60	£37·90	i	£463	£600
e	£1000	£1500	j	£44·20	£66·60

7 The following examples give prices charged by a wholesaler and a retailer for goods. VAT at 15% is *included* in the price of each item. In each example, calculate

(i) the VAT payable on each item
(ii) the amount the retailer pays to the government in VAT. Work to the nearest penny where necessary.

	Wholesaler's price	Retailer's price		Wholesaler's price	Retailer's price
a	£115·00	£172·50	f	£12·60	£18·00
b	£230	£276	g	£27·90	£39·99
c	£1150	£1265	h	£345	£449
d	£30	£45	i	£1800	£2399
e	£74	£99	j	£86	£115

5 Business finances

Exercise 16. Gross and net profit

1 The annual turnover of a small business is £60 000. The table below shows the money spent by the firm.

Cost of materials to make goods	£36 000
Expenses: Labour	7 800
Light and rent	2 900
Rates	3 700
Depreciation of assets	2 800
Other expenses	800

(a) Calculate the *gross profit*, i.e. turnover less the cost of materials.
(b) Calculate the *net profit*, i.e. turnover less the cost of materials and expenses.
(c) Calculate the gross profit per cent on the annual turnover.
(d) Calculate the net profit per cent on the annual turnover.

2 In the following examples calculate the gross profit, the net profit, the gross profit per cent, and the net profit per cent. (Take the percentage to the nearest whole number where necessary.)

	Turnover	Cost of materials	Expenses
a	80 000	48 000	24 000
b	100 000	60 000	25 000
c	30 000	15 000	9 000
d	35 000	21 000	9 000
e	50 000	35 000	4 000
f	67 400	40 000	20 000
g	84 600	50 000	20 000
h	23 200	15 500	6 200
i	12 900	3 200	4 200
j	28 300	19 500	4 800
k	93 400	21 600	62 600
l	72 800	44 500	21 200
m	44 400	28 600	8 900
n	85 100	67 200	5 500
o	94 900	63 000	18 400
p	32 200	21 400	4 700
q	27 500	20 600	4 200
r	47 600	15 100	26 300
s	14 800	2 200	8 300
t	127 000	73 000	35 000

3 An ironmonger has a turnover of £60 000 in one year. His purchases from the wholesaler come to £43 000. His rates for the year are £956, the telephone bill is £263, and the cost of light and heating is £342. He pays his one assistant £2900. What is his net profit per cent for the year (to the nearest whole number)?

4 A factory has a total output of £273 000 per year. Its wage bill amounts to £147 000. Material for its production line costs £50 000. Other expenses come to a total of £31 400. What is the net profit per cent for the year (to the nearest whole number)?

5 A small grocer employs two full-time assistants, one of whom earns £3700, the other £2900. A part-time assistant is paid by the hour and averages a weekly wage of £30. What is his yearly wage bill excluding his own salary?

His sales for the year amount to £73 000. His purchases for the year amount to £49 600. His rates bill is £1500, heating and lighting cost £810, the telephone bill is £180, and he has bad debts amounting to £340. What is his net percentage profit for the year.

6 A firm produces three different models of a washing machine. The table shows the selling price and the number sold each year of each model.

Model	Selling price	Number sold each year
X	£141	670
Y	£163	520
Z	£199	710

The firm employs a staff of 44 who earn an average wage of £63 per week. The rateable value of the factory is £7400 at a rate of 67p in the pound. Heating and lighting the factory costs £64 per week. Other running expenses amount to £10 000 per year. The cost of the materials used to make the washing machines amounts to £100 000 per year. Calculate:

(a) the gross annual profit
(b) the net annual profit
(c) the gross annual profit as a percentage of the total sales (to the nearest whole number)
(d) the net annual profit as a percentage of the total sales (to the nearest whole number).

7 A businessman borrows £150 000 at 15% interest to set up a business. After one year he pays £20 000 of the loan along with the year's interest.

His factory derives its income from the production of 50 articles per week each selling for £120. What is the income for one year where a year is taken to be 50 working weeks?

He employs 40 people for whom the yearly wage bill is £114 600. Heating and lighting for the year is £2500 and the rates bill for the factory is £3400. The cost of the materials for the goods is £80 000 per year. What are the total outgoings per year including the repayment and interest on the loan? What profit is made?

Express this profit as a percentage of the income.

Exercise 17. Partnerships

1 Two men, Able and Black, set up a business together. Mr Able invests capital of £25 000 and Mr Black invests £40 000.

What is the ratio of Able's investment to Black's? The profit at the end of the first year is £9100. This is divided in the ratio of their initial investments.

What share of the profits does each get?

2 Find what share of the profits each partner will get in the examples below.

	Investments			
	Partner A	Partner B	Partner C	Profit
	£	£	£	£
a	3 000	5 000		1 120
b	6 000	8 000		840
c	20 000	45 000		5 850
d	10 000	25 000		3 850
e	13 000	8 000		4 200
f	9 500	3 500		3 900
g	12 000	4 000		2 800
h	4 000	16 000		4 400
i	13 750	13 750		9 050
j	60 500	19 500		16 000
k	3 500	7 000	10 500	3 150
l	5 000	4 000	9 000	6 120
m	15 000	12 000	18 000	6 300
n	28 000	20 000	16 000	8 400
o	3 000	4 000	5 000	1 800
p	6 500	4 000	1 500	2 400
q	40 000	65 000	45 000	37 500
r	12 500	13 000	16 500	10 500
s	100 000	200 000	300 000	60 000
t	2 500	4 000	2 500	2 250

3 The table shows the amount A invested in a partnership, the ratio of A's investment to B's, and the profit the firm made in one year. In each case, find (i) A's share of the profits, (ii) B's share of the profits, (iii) B's capital investment.

	A's investment	A's share : B's share	Profits
a	£ 3 000	1 : 2	£ 1 200
b	6 000	2 : 1	3 000
c	20 000	10 : 2	6 000
d	18 000	4 : 5	9˙000
e	57 000	3 : 2	15 000
f	21 000	3 : 2	7 500
g	9 500	5 : 1	1 500
h	16 000	1 : 3	16 000
i	42 000	3 : 4	21 000
j	47 500	1 : 2	30 000

4 Andrew, Bob and Charles set up as partners in a textile factory. Andrew provided £75 000, Bob £30 000 and Charles £15 000 for the initial capital investment.

Bob acted as manager and received a salary of £7000 in the first year.

Charles was assistant manager and was given a salary of £5000 for the same period.

At the end of the first year the total profits amounted to £28 000. Out of this, Bob's and Charles's salaries were paid and then the *remaining* profits were divided between the three partners in the ratio of the capital each had invested.

(a) What was the total sum provided by the three men to start the business?

(b) What was the profit that was left after Bob's and Charles's salaries had been paid?

(c) How much money did each of the partners get as their share of the *remaining* profits?

(d) How much money did Charles get altogether at the end of the year?

(e) What percentage of the *total* profits did Charles get?

5 Three brothers, Brian, David and Stanley, set up a small fish-processing factory. Brian provided capital of £12 000, David provided £9000, and Stanley £6000. They agreed that 20% of any profits in the first year would be used to expand the business. The *remaining* profits would be shared out in the ratio of the capital investment of each brother. In the first year, the profits amounted to £6750.

(a) What sum of money was taken from the profits to expand the business?

(b) How much money did each of the brothers get?

Exercise 18. Shares

1 Find the market value of the following lots of shares.

  (*a*) 300 £1 shares at £1·25 each
  (*b*) 200 £1 shares at 93p each
  (*c*) 4000 £0·25 shares at £0·27$\frac{1}{2}$ each
  (*d*) 750 £0·50 shares at £0·87 each
  (*e*) 3500 £1 shares at £1·83 each
  (*f*) 2800 £1 shares at 43p each
  (*g*) 7600 £0·25 shares at £0·23 each
  (*h*) 250 £1 shares at £0·74 each
  (*i*) 450 £0·50 shares at £0·57$\frac{1}{2}$ each
  (*j*) 4500 £1 shares at £1·04 each

2 Find the number of shares which can be bought for the following investments at the varying market values.

  (*a*) An investment of £400 in shares at £0·64 each
  (*b*) An investment of £600 in shares at £1·25 each
  (*c*) An investment of £1200 in shares at £1·60 each
  (*d*) An investment of £540 in shares at £1·20 each
  (*e*) An investment of £1148 in shares at £1·40 each
  (*f*) An investment of £8100 in shares at £0·54 each
  (*g*) An investment of £280 in shares at £1·25 each
  (*h*) An investment of £2030 in shares at £1·45 each
  (*i*) An investment of £1767 in shares at £0·95 each
  (*j*) An investment of £896 in shares at £1·12 each

3 Work out the dividends paid each year to the holders of the following shares.

  (*a*) 400 £1 shares paying 8%
  (*b*) 1300 £1 shares paying 12$\frac{1}{2}$%
  (*c*) 2800 £1 shares paying 18%
  (*d*) 900 £0·25 shares paying 7$\frac{1}{2}$%
  (*e*) 4600 £0·50 shares paying 14%

4 Calculate the yield on the following shares. (Work to the nearest whole number.)

  (*a*) £1·00 shares paying 8% and costing £2·00 each
  (*b*) £1·00 shares paying 10% and costing £1·30 each
  (*c*) £1·00 shares paying 7% and costing £0·90 each
  (*d*) £1·00 shares paying 12% and costing £0·73 each
  (*e*) £0·50 shares paying 6% and costing £0·75 each
  (*f*) £0·50 shares paying 5% and costing £0·40 each
  (*g*) £0·25 shares paying 10% and costing £0·30 each
  (*h*) £0·25 shares paying 15% and costing £0·50 each
  (*i*) £1·00 shares paying 12$\frac{1}{2}$% and costing £1·25 each
  (*j*) £1·00 shares paying 7$\frac{1}{2}$% and costing £0·75 each

5 An investor holds 400 £1·00 shares in an engineering company. He sells them when the price per share is £1·21. How much money does he get?
He reinvests this sum in £1·00 shares of a retail company where the price per share is £0·60½. What is the face value of the shares he buys?
If the retail company declares a dividend of 9% what dividend does the investor get? What yield does this represent in the money he invested in the retail company?

6 An investor holds shares in two companies – 300 £1·00 shares at a market value of £1·41 and 200 £1·00 shares at a market value of £0·93½. He sells both lots of shares. How much money does he get?
He invests this money in a company whose £1·00 shares stand at £1·22. What is the face value of the shares he receives?
The company declares a dividend of 14%. How much money does the investor receive? What yield does this represent on his investment in the third company (to 1 decimal place)?

7 The £1 shares in Fish Processors stand at 75p and pay a dividend of 8%. The £1 shares in Market Research Ltd stand at 110p, and pay a dividend of 13%

(a) How much would I pay for shares worth £400 at their face value in

(i) Fish Processors?
(ii) Market Research Ltd?

(b) I invest £660 in *each of these* companies.

(i) How many Fish Processors shares can I buy?
(ii) What dividend do I receive for these?
(iii) How many Market Research Ltd shares can I buy?
(iv) What dividend do I receive for these?
(v) Which is the better investment?

Exercise 19. Stock

1 What will the investor have to pay for the following stocks?

(a) £300 Aircraft 6% stock at 91½
(b) £1000 War Loan 2½% stock at 33
(c) £1400 Associated Magazines 9% stock at 110.
(d) £2300 Blue Line 6½% stock at 63
(e) £4500 American Star 7% stock at 74
(f) £750 Transport 3½% stock at 39
(g) £1200 Ocean Line 8% stock at 105
(h) £9000 British Restaurant 6% stock at 112
(i) £5500 Norcoil 6% stock at 94
(j) £850 Industrial Engineering 8% stock at 102

2 An investor puts the sum of £8505 into each of the following stocks. Find (i) the nominal value, (ii) the annual dividend and (iii) the yield (to 1 decimal place).

(*a*) Gold Securities 8% stock at 70
(*b*) Crosscountry Tyres 6% stock at 60
(*c*) Transport General 5% stock at 120
(*d*) Bank 8% stock at 75
(*e*) Regal Insurance $3\frac{1}{2}$% stock at 90

3 For an investment of £14 586 in each of the following stocks find (i) the nominal value, (ii) the annual dividend and (iii) the yield (to 1 decimal place).

(*a*) Architectural Services 6% stock at 85
(*b*) Peace Loan 8% stock at 110
(*c*) Hi-rise Building $4\frac{1}{2}$% stock at 30
(*d*) Government Gilt 9% stock at 80
(*e*) Noreast $5\frac{1}{2}$% stock at 65

4 An investor holds £7000 of 8% stock at 90. He sells and reinvests the money in 6% stock at 63. What is his change in annual income?

5 An investor holds £850 of $3\frac{1}{2}$% stock at 60. He sells and reinvests the money in 7% stock at 102. What is his change in annual income?

6 £3500 of 7% stock is sold at 73 and £2200 of $3\frac{1}{2}$% stock is sold at 104. The proceeds are reinvested at a rate of $7\frac{1}{2}$%. What is the change in income?

7 £800 of 6% stock is sold at 110 and £500 of $9\frac{1}{2}$% stock is sold at 102. The proceeds are reinvested at a rate of 8%. What is the change in income?

8 I owned £1800 of $4\frac{1}{2}$% Blue Line Stock. I decided to sell when it was at $47\frac{1}{2}$. I invested the proceeds in a chemical works where the £1·00 shares cost £1·50. The chemical works paid a 16% dividend for this year. Calculate:

(*a*) The interest for last year from the $4\frac{1}{2}$% Blue Line Stock
(*b*) The proceeds from the sale of the Blue Line Stock
(*c*) The nominal value of the chemical works shares I bought
(*d*) The dividend I could expect from the chemical works for this year.

6 Foreign exchange

Different countries have their own money systems. The British *unit of currency* is the £, the Italian unit of currency is the lira, the French is the franc. If a businessman wishes to buy goods in another country he must pay in the currency of that country. If a British businessman wishes to buy machinery in France he must buy the francs he requires on the foreign exchange market, usually dealing through a major bank.

The table below shows exchange rates, i.e. how much foreign money one would get for different sums of British money. Use the table to answer the questions in Exercise 20.

Approximate exchange rates

	£1	£5	£10	£50	£100	£1000
Austrian schillings	27·25	136·25	272·50	1362·50	2725	27 250
Belgian francs	58·30	291·50	583·00	2915·00	5830	58 300
Canadian dollars	2·41	12·05	24·10	120·50	241	2410
French francs	8·53	42·65	85·30	426·50	853	8530
Deutsche marks	3·71	18·55	37·10	185·50	371	3710
Italian lire	1670	8350	16 700	83 500	167 000	1 670 000
Spanish pesetas	138·60	693·00	1386·00	6930·00	13 860	138 600
Swiss francs	3·34	16·70	33·40	167·00	334	3340
US dollars	2·01	10·05	20·10	100·50	201	2010

Exercise 20. Exchange by table

1 Change the following amounts into Austrian schillings.

(*a*) £1 (*b*) £50 (*c*) £100 (*d*) £2 (*e*) £60 (*f*) £200

2 Change the following amounts into Italian lire.

(*a*) £1 (*b*) £10 (*c*) £1000 (*d*) £3 (*e*) £40 (*f*) £3000

3 Change the following amounts into US dollars.

(*a*) £5 (*b*) £10 (*c*) £100 (*d*) £15 (*e*) £70 (*f*) £400

4 Change the following amounts into Deutsche marks.

(*a*) £6 (*b*) £300 (*c*) £5000

5 Change the following amounts into French francs.

(a) £25   (b) £90   (c) £10 000

6 Change the following amounts into Canadian dollars.

(a) £250   (b) £750   (c) £2500

7 Change the following amounts into Spanish pesetas.

(a) £20   (b) £200   (c) £2000

8 Change the following amounts into Belgian francs.

(a) £4   (b) £550   (c) £2500

9 Change the following amounts into Swiss francs.

(a) £35   (b) £350   (c) £10 000

Exercise 21. Sterling to other currencies

Calculate these exchanges without using the table.

1 Change £300 to French francs at £1 = 8·53 francs.

2 Change £400 to Swiss francs at £1 = 3·34 francs.

3 Change £900 to Deutsche marks at £1 = 3·71 marks.

4 Change £4000 to Belgian francs at £1 = 58·30 francs.

5 Change £6000 to Austrian schillings at £1 = 27·25 schillings.

6 Change £10 000 to US dollars at £1 = 2·01 dollars.

7 Change £10 000 to Canadian dollars at £1 = 2·41 dollars.

8 Change £25 000 to Portugese escudos at £1 = 94·25 escudos.

9 Change £25 000 to Norwegian kroner at £1 = 10·18 kroner.

10 Change £30 000 to Swedish kronor at £1 = 8·75 kronor.

Exercise 22. Other currencies to sterling

Example

A British businessman exports goods to France. He earns 20 000 francs. What is this in pounds sterling? The rate of exchange is £1 = 8·53 francs.

$$20\ 000\ \text{francs} = £\frac{20000}{8·53}$$

$$= £2344·67$$

1 Change 30 000 Austrian schillings to pounds when £1 = 27·25 schillings.

2 Change 20 000 French francs to pounds when £1 = 8·53 francs.

3 Change 10 000 Deutsche marks to pounds when £1 = 3·71 marks.

4 Change 5000 Belgian francs to pounds when £1 = 58·30 francs.

5 Change 40 000 Spanish pesetas to pounds when £1 = 138·60 pesetas.

6 Change 20 000 US dollars to pounds when £1 = 2·01 dollars.

7 Change 60 000 Italian lire to pounds when £1 = 1670 lire.

8 Change 40 000 Canadian dollars to pounds when £1 = 2·41 dollars.

9 Change 15 000 Swiss francs to pounds when £1 = 3·34 francs.

10 Change 30 000 Portugese escudos to pounds when £1 = 94·25 escudos.

Exercise 23. Problems

1 A British businessman buys refrigerators from an Italian firm. The value of the order is £2000. The exchange rate is £1 = 1670 lire. How many lire will the businessman have to pay?

2 Spare parts for cars are bought from France. The exchange rate is £1 = 8·53 francs. The total bill is £1700. What is the cost in francs?

3 A British firm buys goods in Belgium. The exchange rate is £1 = 58·30 francs. If the cost of the goods is £3500, what is the cost in francs?

4 A holiday villa in Portugal is to cost £240 per week. The exchange rate is £1 = 94·25 escudos. What is the cost in escudos of 3 weeks in the villa?

5 If Norwegian skiing equipment is to cost £1200, what will its cost be in Norwegian kroner? £1 = 10·18 kroner.

6 The cost to a British businessman of goods bought in Austria is 38 150 schillings. The exchange rate is £1 = 27·25 schillings. How much is this in pounds?

7 Goods bought in Sweden cost 21 875 kronor. The exchange rate is £1 = 8·75 kronor. What is the cost of the goods in pounds?

8 A charter return flight to New York is to cost 422 dollars to the nearest dollar. £1 = 2·01 dollars. What is the cost of the flight to the nearest pound?

9 Presents bought in Canada by a holidaymaker amount to 221·72 dollars. The rate of exchange is £1 = 2·41 dollars. What did the holidaymaker pay in pounds for these presents?

10 A British dealer buys 2000 digital watches and 300 mechanical watches in Switzerland. The cost of each digital watch is 30·06 francs; the cost of each mechanical watch is 56·78 francs. The rate of exchange is £1 = 3·34 francs. Find the total cost to the dealer in pounds.

Exercise 24. Extended problems

1 The rate of exchange for Spanish currency is 138·60 pesetas to the £1. A person going to Spain on business changed £240 into pesetas.

(*a*) How many pesetas did he receive?
(*b*) He spent 27 440 pesetas and changed the remainder into £ on his return. How much did he receive, in £, on his return?

2 A businessman goes on a sales visit to France. He is allowed expenses of £250. The exchange rate for buying francs is £1 = 8·53 francs and he changes his £250 into francs at this rate. In fact the expenses he incurs amount to 1940 francs and he changes all his remaining francs back into sterling at a rate of £1 = 8·58 francs.

(*a*) How many francs did he get for his £250?
(*b*) How many pounds did he use to buy francs for his expenses?
(*c*) How many pounds did he receive on returning to Britain?

3 A grocer wins a national competition for the best window display. The prize is a package holiday for two in Toronto. The total cost of the package is £930 plus £250 in spending money for each person.

(*a*) What is the total value of the prize?
(*b*) The rate of exchange is 2·41 Canadian dollars to £1. What is the cost of the package in dollars excluding the spending money?
(*c*) They convert £380 into Canadian dollars. How many dollars do they get?
(*d*) The remaining spending money is changed into US dollars for a short visit there. 2·01 dollars = £1. How many US dollars do they get?

4 The rate of exchange is £1 = 58·30 Belgium francs but £1 = 8·53 French francs.

(*a*) How many Belgium francs would a person get for £70?
(*b*) How many French francs would a person get for £70?
(*c*) A travel firm changes 430 Belgium francs for a day trip. How much is this in sterling (to the nearest penny)?
(*d*) How many French francs would a person get for 1166 Belgium francs (to the nearest franc)?

5 In Switzerland £1 = 3·34 francs.
In Austria £1 = 27·25 schillings

(*a*) How many Swiss francs would a person get for 109 Austrian schillings?
(*b*) How many schillings would a person get for 668 Swiss francs?
(*c*) How many francs would a person get for 3000 schillings?
(*d*) How many schillings would a person get for 1000 francs?

6 This graph is a conversion graph from £ to Canadian dollars.

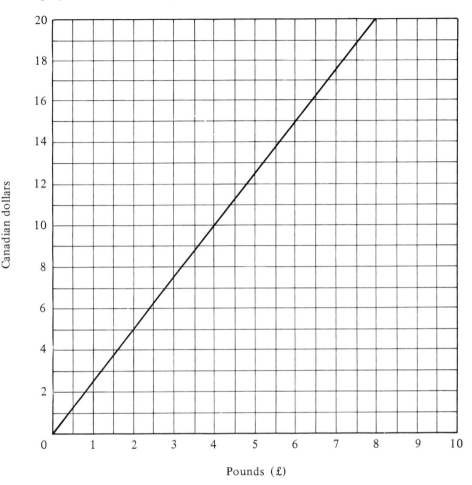

Pounds (£)

Use the graph to answer the following questions.

(a) What is the value of £1 in Canadian dollars?
(b) How many Canadian dollars would you receive for £6·50?
(c) What sum of money in sterling would you get for (i) 10 Canadian dollars, (ii) 17 Canadian dollars?

7 Last year, the rate of exchange in Germany was £1 equal to 3·50 Deutsche marks.
Draw a conversion graph for small sums of money from British to German currency using this rate of exchange. For your scale use 4 cm to represent £1 and 1 cm to represent 1 Deutsche mark. Make both axes 12 cm long. Use your graph to find the following.

(a) £2·25 in marks   (b) 50p in marks   (c) 8 marks in £
(d) 10·5 marks in £

Answers

Exercise 1

1 (a) £104   (b) £1125   (c) £336   (d) £504   (e) £3120   (f) £262·50
   (g) £57·75   (h) £1148·85   (i) £2025   (j) £10 773·75   (k) £1001·70
   (l) £1109·25   (m) £270   (n) £236·80   (o) £453·60   (p) £2285·94
   (q) £1654·81   (r) £1049·75   (s) £6201·56   (t) £1984·68

2 (a) 7% p.a.   (b) $5\frac{1}{2}$% p.a.   (c) $8\frac{1}{4}$% p.a.   (d) $11\frac{1}{2}$% p.a.   (e) 9% p.a.
   (f) $13\frac{1}{2}$% p.a.

3 (a) 4   (b) 3   (c) $4\frac{1}{2}$ (d) $2\frac{3}{4}$ (e) 4   (f) $2\frac{1}{2}$

4 (a) £1500   (b) £20 000   (c) £3200   (d) £650   (e) £1750   (f) £900

Exercise 2

1 £220   2 6·5% per annum   3 £432   4 £650   5 4 years
6 £1600   7 £1500   8 (a) £28   (b) 6 years   (c) £63   (d) £87·50

Exercise 3

1 (a) £50   (b) £1050   (c) £52·50   (d) £1102·50   (e) £55·13

2 £112·49   3 £998·40   4 £2072·05   5 £2015·54   6 £29 282

7 £37·79

8 (a) £75 000   (b) £78 812·50   (c) £3812·50   (d) 5·254%

9 (a) £404·93   (b) £8267·52   (c) £10 000   (d) £2000   (e) £30 000
   (f) 13% per annum   (g) £400   (h) £6000

Exercise 4

1 (a) £100   (b) £200   (c) £250   (d) £500

2 (a) £1125   (b) £1495·08   (c) £840   (d) £1800

3 (a) £210, £89·45   (b) £350, £58·34   (c) £560, £40·84   (d) £700, £35

4 24 months

5 (a) 48 months   (b) 30 months   (c) 60 months   (d) 12 months   (e) 18 months

6 (a) £1350   (b) £5850   (c) £162·50

Exercise 5

1 (a) £1800   (b) £160   (c) £110·50   (d) £240·63   (e) £481   (f) £77·40
   (g) £799   (h) £1155   (i) £288   (j) £149·40

2 (a) £130   (b) £39   (c) £91

3 (a) £136·50   (b) £147   (c) £157·50   (d) £262·50   (e) £703·50   (f) £1709·40
   (g) £1264·20   (h) £94·43   (i) £283·22   (j) £550·62

4 (a) £6600   (b) £26 400   (c) £274·56   (d) £82 368

5 (a) £6650, £8680, £9590, £13 860   (b) £67·83, £88·54, £97·82, £141·37

Exercise 6

1 (a) £2·69 (b) £1·84 (c) £16·25 (d) £4·17 (e) £9 (f) £4·50
  (g) £24·50 (h) £4·74 (i) £58·32 (j) £3·45

2 (a) £2·66 (b) £3·17 (c) £3·47 (d) £6·06 (e) £4·88 (f) £6·75
  (g) £10·39 (h) £10·04 (i) £26·66 (j) £25·85

3 (a) £292·40 (b) £34·40 (c) £77·40 (d) £16·13 (e) £94·72

4 (a) £89 (b) £4·50 (c) 8·26% (d) 11·01%

5 (a) £319·20 (b) £439·68 (c) £135 (d) 3 years

6 (a) £747 (b) £48 (c) 9·16% (d) £226·92

Exercise 7

1 (a) £9600 (b) £36 000 (c) £29 520 (d) £68 000 (e) £8750
  (f) £14 720 (g) £14 820 (h) £53 760 (i) £15 470 (j) £18 060

2 (a) £7680 (b) £27 000 (c) £12 283 (d) £6030 (e) £25 392
  (f) £21 206 (g) £8044 (h) £60 286 (i) £44 977 (j) £31 063

3 Large car (a) £1240 (b) £744 (c) £3794
  Middle car (a) £920 (b) £552 (c) £2815
  Small car (a) £560 (b) £336 (c) £1714

4 (a) £285 (b) £228 5 83·6%

Exercise 8

1 (a) £1·50 (b) £1·50, £60, £0·90, £40, £0·60, £20, £0·30
  (c) (i) £103·30, (ii) £3·30, (iii) 3·3% (d) £40

2 (a) £9, £400, £8, £320, £6·40, £150, £3 (b) £153 (c) £26·40

Exercise 9

1 (a) 40p (b) £1·20 (c) £2·17$\frac{1}{2}$ (d) £8·52$\frac{1}{2}$ (e) £93·62$\frac{1}{2}$
  (f) £155·75 (g) £293·15 (h) £512·40 (i) £802·75 (j) £218

2 (a) £1·76 (b) £4·84 (c) £8·47 (d) £21·78 (e) £27·17
  (f) £40·48 (g) £60·94 (h) £280·94 (i) £760·10 (j) £5148

3 (a) 60p (b) £1·20 (c) £1·58 (d) £3·38 (e) £6·66
  (f) £10·10 (g) £15·42 (h) £25·72 (i) £116·66 (j) £832

4 (a) £0·81 (b) £1·25 (c) £1·53 (d) £2·97 (e) £3·80
  (f) £5·40 (g) £17·73 (h) £54 (i) £108 (j) £810

5 (a) £9·72 (b) £10·80 (c) £15·09 (d) £0·90 (e) £7·83
  (f) £23·65 (g) £35·09 (h) £108·50

6 (a) £60 (b) £200 (c) £360 (d) £159 (e) £373·80
  (f) £73·50 (g) £229·50 (h) £671·50

7 (a) £26·40 (b) £66 (c) £132 (d) £18·48 (e) £47·52
  (f) £142·56 (g) £121·44 (h) £33 (i) £83·16 (j) £147·84

8 (a) £37·88 (b) £50 (c) £100 (d) £18·94 (e) £30·30
  (f) £90·91 (g) £25·76 (h) £54·55 (i) £65·91 (j) £108·33

Exercise 10

1 £40·35, 17·65%   **2** £39·99, 25%   **3** £105, 16·28%

4 £179·95, £44·95, 33%   **5** £299·86, £89·86, 43%

6 £179·95, £62·95, 54%   **7** £186·63, £93·32, 50%

8 £39·49, £13·16, 12·24%   **9** £127, £54·45, £163·31, 22·23%

10 £3·08   **11** £8·28

12 (*a*) £590 500 (*b*) £469 500 (*c*) 97 200 (*d*) 82 800 (*e*) £826 200
(*f*) £621 000 (*g*) £356 700 (*h*) £30 500

13 £654 750, £290 750

14

	Cost price	Normal selling price	Sale price
a	£4·20	£6·04	£5·43
b	£4·59	£6·60	£5·94
c	£6·57	£9·44	£8·50
d	£10·25	£14·73	£13·26
e	£10·26	£14·75	£13·28
f	£15·45	£22·21	£19·99

15

	Cost price	Normal selling price	Sale price
a	£2·30	£3·44	£3·16
b	£3·10	£4·64	£4·27
c	£5·08	£7·60	£6·99
d	£11·43	£17·09	£15·72
e	£10·45	£15·63	£14·38
f	£14·69	£21·96	£20·20

16

	Cost price	Normal selling price	Sale price
a	£3·50	£5·43	£4·35
b	£6·18	£9·60	£7·68
c	£9·98	£15·50	£12·40
d	£2·25	£3·49	£2·79
e	£5·73	£8·90	£7·12
f	£19·73	£30·63	£24·50

17 (*a*) £22 400 (*b*) £7210 (*c*) 32% (*d*) £222·08 (*e*) £255·39

18 (*a*) 84% (*b*) £245·63 (*c*) £19·92

19 £3·28, 27%

20 £102·13, £116·72, £14·59, 14%

Exercise 11

1 (*a*) £60·50 (*b*) £57·20 (*c*) £63·80 (*d*) £55·55 (*e*) £52·25 (*f*) £62·50
(*g*) £53·13 (*h*) £70 (*i*) £60·63 (*j*) £55 (*k*) £58·73 (*l*) £60·75
(*m*) £72·90 (*n*) £62·78 (*o*) £54·68 (*p*) £98 (*q*) £103·25 (*r*) £108·50
(*s*) £71·75 (*t*) £93·50 (*u*) £115·50 (*v*) £132 (*w*) £112·20 (*x*) £108·90
(*y*) £105·60 (*z*) £38·54 (*A*) £39·95 (*B*) £44·18 (*C*) £52·64 (*D*) £37·13

2 (a) £6·96   (b) £34·80   (c) £118·32   (d) £49·59   (e) £164·43, £54·81
3 (a) £20·09   (b) £100·45   (c) £40·18
4 (a) 6   (b) 184   (c) 97   (d) 9   (e) 31
5 (a) £41·32$\frac{1}{2}$ (b) £45·24   (c) £46·98   (d) £42·19$\frac{1}{2}$ (e) £37·84$\frac{1}{2}$
6 £4221   7 (a) £101·06   (b) £5659·36
8 (a) £60·80   (b) £8   (c) £90
9 (a) £50·80   (b) £74·29$\frac{1}{2}$ (c) 5 hours   (d) 12 hours
10 (a)

	Basic hours worked	Hours of overtime worked
Monday	8	
Tuesday	8	
Wednesday	7$\frac{1}{4}$	
Thursday	8	2$\frac{1}{2}$
Friday	7$\frac{3}{4}$	
Saturday		4
Sunday		4$\frac{1}{4}$

(b) £128·15   (c) £101·75   (d) £6663·80, £5291   (e) £1372·80

Exercise 12

1 (a) £33·20   (b) £40·20   (c) £38·40   (d) £34·60   (e) £37·05
 (f) £44   (g) £41·40   (h) £40·60   (i) £45·80   (j) £46·20
2 £4000, £333·33   3 £5400, £450
4 (a) £4380, £4800, £4950   (b) £365, £400, £412·50   (c) £14 130
 (d) £4710   (e) £174 270
5 £108 000   6 7%   7 6%

Exercise 13

1 (a) £201   (b) £332·99   (c) £428·80   (d) £7597·80   (e) £38 123
 (f) £76 380
2 (a) £2400   (b) £3800   (c) £14 000   (d) £363 430   (e) £815 680
 (f) £1 103 474·32
3 (a) 53p   (b) 49p   (c) 56p   (d) 64p   (e) 69p
4 £57 755·52, £60 162, 4%
5 (a) £73 576 050   (b) £9 883 350   (c) £229·81   (d) £273
6 (a) 60 p   (b) 55p
 (c) (i) £29 937·60   (ii) £22 317·12, £7620·48   (iii) £32 114·88
 (d) (i) £39 144·60   (ii) £29 180·52, £9964·08   (e) (i) £457   (ii) £36·56

Exercise 14

1 (a) £280   (b) £840   (c) £140   (d) £770   (e) £490   (f) £80   (g) £760
 (h) £530   (i) £670   (f) £220
2 (a) £1000   (b) £2000   (c) £360   (d) £1430   (e) £2500   (f) £2250
 (g) £750   (h) £1960   (i) £2460   (j) £1360

3 (a) £60   (b) £150   (c) £300   (d) £450   (e) £810   (f) £1020   (g) £1701
  (h) £390   (i) £1035   (j) £1377

4 (a) £165   (b) £330   (c) £1089   (d) £221   (e) £135   (f) £508   (g) £1907
  (h) £1551   (i) £92   (j) £117

5 (a) (i) £1380   (ii) £414   (iii) £3586
  (b) (i) £582   (ii) £3978   (iii) £392   (iv) 9·8%

6 £97·53

8 (a) £830   (b) £755   (c) £75   (d) 1·5%

9 (a) £22·66, £29·06, £49·26, £63·05, £66·99, £76·35
  (b) £74·63, £100, £132·84, £235·83, £177·62, £86·57
  (c) £76·12, £64·18, £107·47, £164·18, £231·35, £144·78

Exercise 15

1 (a) £53·48   (b) £90·85   (c) £57·44   (d) £44·79   (e) £366·85   (f) £4·03
  (g) £84·18   (h) £74·29   (i) £26·85   (j) £574·94   (k) £4·49   (l) £12·02
  (m) £126·85   (n) £14·95   (o) £16·91   (p) £21·39   (q) £14·55
  (r) £8·97   (s) £21·51   (t) £16·33

2 (a) £5·86   (b) £0·96   (c) £6·21   (d) £64·70   (e) £107·74   (f) £214·35
  (g) £551·30   (h) £15·57   (i) £73·30   (j) £17·30   (k) £31·65
  (l) £185·04   (m) £136·26   (n) £4·04   (o) £0·87   (p) £299·04
  (q) £70·96   (r) £16·70   (s) £15·48   (t) £95·96

3 £8·80, £9·90, £11·38$\frac{1}{2}$ 4 £6·45, £73·60, £3·15   5 £45, £67·50, £22·50

6 (a) (i) £0·51, £0·78   (ii) £0·27   (b) (i) £1·18$\frac{1}{2}$, £1·56   (ii) £0·37$\frac{1}{2}$
  (c) (i) £2·74$\frac{1}{2}$, £3·61$\frac{1}{2}$ (ii) £0·87   (d) (i) £4·14, £5·68$\frac{1}{2}$ (ii) £1·54$\frac{1}{2}$
  (e) (i) £150, £225   (ii) £75   (f) (i) £51·45, £70·35   (ii) £18·90
  (g) (i) £16·80, £20·25   (ii) £3·45   (h) (i) £14·85, £19·35   (ii) £4·50
  (i) (i) £69·45, £90   (ii) £20·55   (j) (i) £6·63, £9·99   (ii) £3·36

7 (a) (i) £15, £22·50   (ii) £7·50   (b) (i) £30, £36   (ii) £6
  (c) (i) £150, £165   (ii) £15   (d) (i) £3·91, £5·87   (ii) £1·96
  (e) (i) £9·65, £12·91   (ii) £3·26   (f) (i) £1·64, £2·35   (ii) £0·71
  (g) (i) £3·64, £5·22   (ii) £1·58   (h) (i) £45, £58·57   (ii) £13·57
  (i) (i) £234·78, £312·91   (ii) £78·13   (j) (i) £11·22, £15   (ii) £3·78

Exercise 16

1 (a) £24 000   (b) £6000   (c) 40%   (d) 10%

2 (a) £32 000, £8000, 40%, 10%   (b) £40 000, £15 000, 40%, 15%
  (c) £15 000, £6000, 50%, 20%   (d) £14 000, £5000, 40%, 14%
  (e) £15 000, £11 000, 30%, 22%   (f) £27 400, £7400, 41%, 11%
  (g) £34 600, £14 600, 41%, 17%   (h) £7700, £1500, 33%, 6%
  (i) £9700, £5500, 75%, 43%   (j) £8800, £4000, 31%, 14%
  (k) £71 800, £9200, 77%, 10%   (l) £28 300, £7100, 39%, 10%
  (m) £15 800, £6900, 36%, 16%   (n) £17 900, £12 400, 21%, 15%
  (o) £31 900, £13 500, 34%, 14%   (p) £10 800, £6100, 34%, 19%
  (q) £6900, £2700, 25%, 10%   (r) £32 500, £6200, 68%, 13%
  (s) £12 600, £4300, 85%, 29%   (t) £54 000, £19 000, 43%, 15%

3 21%   **4** 16%   **5** £8160, 17%

6 (a) £220 520   (b) £58 090   (c) 69%   (d) 18%

7 £300 000, £243 000, £57 000, 19%

Exercise 17

1 5 : 8; £3500; £5600

2 (a) £420, £700   (b) £360, £480   (c) £1800, £4050   (d) £1100, £2750
(e) £2600, £1600   (f) £2850, £1050   (g) £2100, £700   (h) £880, £3520
(i) £4525, £4525   (j) £12 100, £3900   (k) £525, £1050, £1575
(l) £1700, £1360, £3060   (m) £2100, £1680, £2520   (n) £3675, £2625, £2100
(o) £450, £600, £750   (p) £1300, £800, £300   (q) £10 000, £16 250, £11 250
(r) £3125, £3250, £4125   (s) £10 000, £20 000, £30 000
(t) £625, £1000, £625

3 (a) £400, £800, £6000   (b) £2000, £1000, £3000   (c) £5000, £1000, £4000
(d) £4000, £5000, £22 500   (e) £9000, £6000, £38 000
(f) £4500, £3000, £14 000   (g) £1250, £250, £1900
(h) £4000, £12 000, £48 000   (l) £9000, £12 000, £56 000
(j) £10 000, £20 000, £95 000

4 (a) £120 000   (b) £16 000
(c) Andrew £10 000, Bob £4000, Charles £2000   (d) £7000   (e) 25%

5 (a) £1350   (b) £2400, £1800, £1200

Exercise 18

1 (a) £375   (b) £186   (c) £1100   (d) £652·50   (e) £6405   (f) £1204
(g) £1748   (h) £185   (i) £258·75   (j) £4680

2 (a) 625   (b) 480   (c) 750   (d) 450   (e) 820   (f) 15 000   (g) 224
(h) 1400   (i) 1860   (j) 800

3 (a) £32   (b) £162·50   (c) £504   (d) £16·87$\frac{1}{2}$ (e) £322

4 (a) 4%   (b) 8%   (c) 8%   (d) 16%   (e) 4%   (f) 6%   (g) 8%   (h) 8%
(i) 10%   (j) 10%

5 £484, £800, £72, 15%

6 £610, £500, £70, 11·5%

7 (a) (i) £300   (ii) £440
(b) (i) £880   (ii) £70·40   (iii) £600   (iv) £78   (v) Market Research Ltd

Exercise 19

1 (a) £274·50   (b) £330   (c) £1540   (d) £1449   (e) £3330   (f) £292·50
(g) £1260   (h) £10 080   (i) £5170   (j) £867

2 (a) £12 150, £972, 11·4%   (b) £14 175, £850·50, 10%
(c) £7087·50, £354·37$\frac{1}{2}$, 4·2%   (d) £11 340, £907·20, 10·7%
(e) £9450, £330·75, 3·9%

3 (a) £17 160, £1029·60, 7·1%   (b) £13 260, £1060·80, 7·3%
(c) £48 620, £2187·90, 15%   (d) £18 232·50, £1640·92$\frac{1}{2}$, 11·3%
(e) £22 440, £1234·20, 8·5%

4 £40   **5** £5·25   **6** £41·22½   **7** £15·70

8 (*a*) £81   (*b*) £855   (*c*) £570   (*d*) £91·20

Exercise 20

1 (*a*) 27·25   (*b*) 1362·50   (*c*) 2725·00   (*d*) 54·50   (*e*) 1635·00
(*f*) 5450·00

2 (*a*) 1670   (*b*) 16 700   (*c*) 1 670 000   (*d*) 5010   (*e*) 66 800
(*f*) 5 010 000

3 (*a*) 10·05   (*b*) 20·10   (*c*) 201   (*d*) 30·15   (*e*) 140·70
(*f*) 804

4 (*a*) 22·26   (*b*) 1113   (*c*) 18550

5 (*a*) 213·25   (*b*) 767·70   (*c*) 85 300

6 (*a*) 602·50   (*b*) 1807·50   (*c*) 6025

7 (*a*) 2772   (*b*) 27 720   (*c*) 277 200

8 (*a*) 233·20   (*b*) 32 065   (*c*) 145 750

9 (*a*) 116·9   (*b*) 1169   (*c*) 33 400

Exercise 21

1 2559   **2** 1336   **3** 3339   **4** 233 200   **5** 163 500

6 20 100   **7** 24 100   **8** 2 356 250   **9** 254 500   **10** 262 500

Exercise 22

1 £1100·92   **2** £2344·67   **3** £2695·42   **4** £85·76   **5** £288·60

6 £9950·25   **7** £359·28   **8** £16 597·51   **9** £4491·02   **10** £318·30

Exercise 23

1 3 340 000 lire   **2** 14 501 francs   **3** 204 050 francs   **4** 67 860 escudos

5 12 216 kronor   **6** £1400   **7** £2500   **8** £210   **9** £92

10 £23 100

Exercise 24

1 (*a*) 33264   (*b*) £42·02

2 (*a*) 2132·50   (*b*) £227·43   (*c*) £22·44

3 (*a*) £1430   (*b*) 2241·30   (*c*) £915·80   (*d*) 241·20

4 (*a*) 4081   (*b*) 597·1   (*c*) £7·38   (*d*) 171

5 (*a*) 13·36   (*b*) 5450   (*c*) 367·71   (*d*) 8158·68

6 (*a*) 2·5   (*b*) 16·3   (*c*) (i) £4   (ii) £6·80

7 (*a*) 7·9   (*b*) 1·75   (*c*) £2·29   (*d*) £3